WE ARE HERE

A MEMOIR

A YOUNG MAN'S VISITATIONS
FROM ADVANCED BEINGS

BY ANDRE LAWRENCE

We Are Here - The Memoir

Copyright © 2020 Andre Lawrence

ISBN 978-1-7362812-0-8

THAT YE,
BEING ROOTED
AND GROUNDED
IN LOVE,
MAY BE ABLE TO
COMPREHEND
WITH ALL SAINTS
WHAT IS
THE BREADTH,
AND LENGTH,
AND DEPTH,
AND HEIGHT

Paul the Apostle – Ephesians 3:17

Have you ever wondered if there was anyone else besides us in the universe? My whole life, I've been fascinated with the possibility of there being something more advanced among us; teleportation or holographic projections, for example. Though it is an outrageous thought, there could very well be a best-kept secret from the general population of this world. Multiple locations on this Earth we find ancient pyramids engraved with coded messages, such as Egypt, Iraq, Sudan, Indonesia, Peru, Guatemala and Mexico. Their placements are very similar to microchip boards we find in computers today! We also find monolithic sculptures and crypted wisdom where various ancient civilizations such as the Mayans, Greeks, Native Americans and Celtic Druids each depicting almost identical creatures or "gods," leaving us with a feeling that there are more unexplained phenomena out there. Could there be other beings more advanced than humanity that existed here before us? Will we ever meet beings of other worlds in this lifetime? At some point we have all entertained this thought and not get much further than that.

At the sprouting age of five, I was introduced to the concept of Catholicism. I didn't understand the whole ritual and purpose of it, but I went along with the traditions and holidays. I was never baptized so I felt like I had the freedom to explore and learn about other belief systems. To me, religion is whatever I believed to be real and felt in my heart; it's about making the right decisions and being truthful. Shouldn't we have the power within our own minds and hearts to make the right choices? Buddhism has been more of my understanding and reminder that we must always be conscious of ourselves while being respectful to all other forms of life around us, with awareness and gratitude.

Throughout my childhood, every so often my sister and I would get "brush-downs" from my incredibly talented and wise mom and grandma. They were considered to be Light Workers, or healers, from Hungary. Brush-downs, also known as clearings, were like massages that barely touched the body. Their hands would hover a few inches away from the head, then slowly move down the torso, legs, and finally over the heel or toes, clearing any bad energy away by pulling an electromagnetic force with the hands. It was similar to Reiki. The concept was intriguing to me; how the body moves or creates energy, and I would love to know more about this kind of knowledge and practice. Naturally, I turned to science, space, astronomy and astrology for answers. Most importantly, I turned to spirituality. I was questioning everything. My dad always said, "Never be afraid to ask questions." Since I am more of a creative and artistic person, my mind always questioned what is real and what isn't. How far could my imagination take me, testing the boundaries of reality.

There are facts to life that we live with but rarely question, such as the sun, the moon and the stars. They are consistently in motion around us for centuries and are a major part of why we are here, but we don't give them a second thought. This is where I would look deeper. Is there something more convincing or real than God and spirituality as an answer and reason for human existence? Spirituality is our connection to everything else out there – that's here, now. Who and what exactly are we to deserve such an honor of experiencing this physical life, roaming around in an intricately designed vessel, absorbing light through our eyes and into our complex brains, sharing this beautiful planet with each other. There must be more to than just existing; there has to be. Mathematical consistencies and geometrical patterns are an ever flowing force of nature within and everywhere around us. Humans have come so far in such a short flash of a few centuries. Throughout history, brilliant minds have gathered a wealth of knowledge from years of studies, tests and research and yet there is much more to discover within us. Where do we go from here?

In the year 2011, I turned 29. My attention was turning more to the skies, wondering how great it would be to fly and what is really out there in the vast universe. I have always been drawn to the stars and constellations, especially on dark and crisp nights, and I couldn't help but admire how many stars there were and the patterns they created, like a guiding map. I stood perfectly and quietly still, focusing intently on little clusters of stars and the patterns they created as they sat there static, twinkling softly. Just imagine how long these patterns of light have been around. Witnessed by many generations of mankind, while we are just a mere blink on the timeline.

Every once in a while, I would see faint or dim lights that may look like stars, drifting carefully through the air heading in a random direction. Many of us have noticed these and determined it could be one of a million things, like a satellite or airplane. Then we just forget about them. I know an airplane when I see one, but they don't move or blink like these little lights do. At brief moments, their light would brighten, and then dim or completely fade out. It was a relief to know others saw them as well when I pointed them out, only so I knew it wasn't just my eyes playing games with me. The typical reactions to them were that they're satellites, weather balloons, or shooting stars, but not many people thought they were much else. The more that I tried to ignore them, the more I would notice these stars going in all sorts of directions, unlike any usual flight patterns of airplanes or satellites.

On a typical evening, about a half hour after dinner, I would do a quick workout and then lay at the end of the driveway of my mom's house. I would look up and see these stars seeming to fly by, but closer than usual. Why do I keep seeing these things? What are they? I thought, "This star-like object is drifting across what seems so far away in space, but it could very well be just within a thousand meters." I reached out to it in a whispered thought, with true intent to communicate with it, "Where are you going and what are you doing?" How will I find the answers I'm looking for, unless I ask the right questions? What are the questions to the answers I need to know: "Who am I, Why am I here?" Elaborating,

I continued, "What am I to do on this earth, with all the skills, talents and wisdom I've earned up until now? I would love to help make the world a better place, so what is it that I can do in order to be where I need to be? I am ready and willing to make a difference in this world for all of humanity and for the Earth. There are many changes happening and I am ready and willing to receive your guidance. Please, show me a sign." With all of my heart and soul, I meant every word of this. After that night, I never expected anything to happen. If something did, how would I know? The next few months were about to reveal a lot more to me than I could've ever imagined.

On February 16, 2011, I saw an unusual sighting unlike the slowly drifting stars that I'd been seeing before. I had made a stop in Stamford, Connecticut in the parking lot of Cove Beach to enjoy a late lunch. While sitting in my car, I was just about to take a bite of my sandwich when I looked up and saw, in the near distance to the north, a round floating object slowly hovering its way out to the Long Island Sound, steadily drifting up or downwards, varying in altitude. There was one large circle with another round circle inside it rotating and turning, kind of like a gyroscope. Luckily, I had my Nikon D90 camera with a zoom lens in my trunk so I used that to capture a few photos.

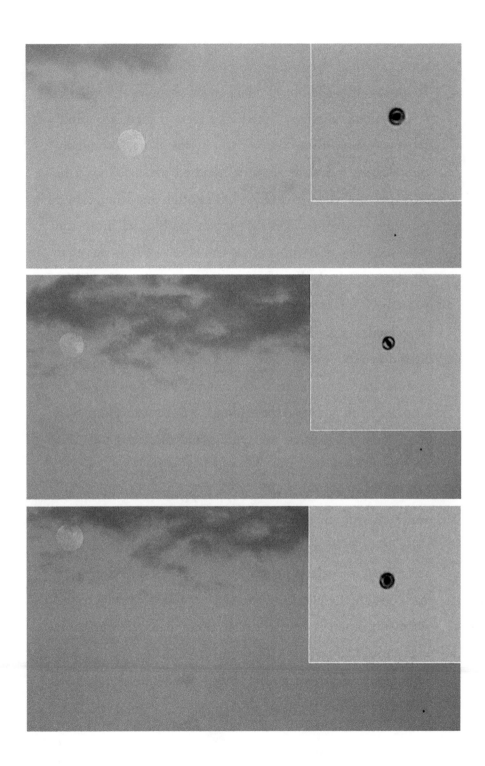

It appears that the object changed its shape in the middle. It was unlike anything I had ever seen, it blew my mind! As it wavered off into the horizon by the shore, it suddenly disappeared and then about a hundred geese followed in its path. Just when it couldn't have gotten any stranger, for a couple of seconds, hail the size of marbles fell from the sky on a clear day with only a few clouds. What was that all about? I looked up and there was a misty, cloud-like substance, but mostly there was blue sky! Just after that happened, I stopped by my good friend Zhibez's house. He lived down the street and the first thing he said was: "Dre did you see that hail coming down not too long ago?" I said, "I sure did! You gotta see these photos I just took right before it… Look!" I grabbed my camera and displayed them on the digital screen for him. He was just as stunned as I was. We both knew there were changes happening and the world was constantly in motion, growing, evolving. We reminded each other to be ready for anything and to evolve ourselves in any way we can. Who knows what else could happen? I had to share the photos of this object on social media to see what everyone else thought of it. Some have tried to say it was photo-shopped or a fake but I have the original photos all together. And I'm sure there was someone else out there that day who saw the same thing.

Around the same time as these strange sightings began, I was introduced to a Lutheran church in Norwalk by the pastor, who was also the pitcher on a softball team I played on at the time. I began attending St. Peter's Lutheran Church regularly and I felt blessed to be able to sit with the pastors' family during services. Many of my good friends and old teachers happened to be members of this church as well! Everyone was truly kind and caring, and the energy was always so powerful and so it piqued my interest. I felt I was making a strong spiritual connection there. Perhaps I would make a stronger connection with these things I have been seeing as of late, too. Most importantly to note, Pastor Beinke and his family were there for me during difficult times and were always willing to lend a hand, to listen with a true and supporting kindness in their hearts.

On the evening of March 18, 2011, the Beinke's invited me over to watch a movie and I gladly accepted the invite. It was an old classic movie called The Great Race. It ended just before 10 o'clock at night, so it was about time to go home. Leaving their house with a euphoric feeling we tended to have after a movie, I walked out to my car in the dark parking lot (they lived right next to the church), I heard what sounded like a ferocious flame or fireball, kind of like a big sparkler igniting. Ffpphttphhfffffphhttfftfttphhff. I looked overhead and, coming from the right, was an amber-colored fireball just skimming over the tops of the trees. It was making its way directly over my head and, as it slowed directly above me, it made a sharp turn right, then another turn left and continued straight. "What the?!?" I wondered. I thought about walking back into the house to let everyone know what I just saw. But there was nothing to show them, no proof. Not wanting to create a scene, I kept that moment to myself and went home in amazement.

This flying fireball ignited my curiosity to look up records of any other similar sightings. I signed up on UFO chat forums to discuss these unfamiliar objects, and, even when I shared the photos I took, I guess the forum members thought I was making it all up and didn't accept that I was coming from a place of truth and honesty. In one of the chat forums, someone reached out to me to share experiences and information. They were genuinely interested so we remained in contact on social media. Chatting with this person gave me a bit of reassurance and the strength to continue my search for answers. After all, I kept seeing more of these randomly drifting, mysterious stars each night so I had to continue questioning.

Only four days after seeing this fireball, on March 22, 2011, I was hanging out with some friends in Stamford and, of course, I had to let them know about these flying things. I had to see what they thought or felt about them. When you haven't seen something in person with your own eyes, it's really hard to visualize it and draw a conclusion as to what might be happening. But they all said that, as long as there's no danger or threat, then it's all good, right? Love and respect for all creations of life! They gave me hope that there should be nothing to worry about, and to accept whatever is happening to me (if something was happening at all).

I started to make my way home around 11 O'clock that night. I arrived in Danbury, pulling onto my street just before midnight. Taking my time up the street, as the stars were magnificent this night, I saw a star, brighter than a normal one and it looked as if it were descending. Perhaps it was the car moving down the hill creating this illusion, so I quickly pulled to the side of the road, parked and got out to really look. Sure enough, it was descending very quietly and gently. I couldn't tell how far or close it was. Quickly, I thought to flash the bright lights on my car to try and communicate with it.

I was reaching into the car window, flashing steadily, One... Two... Three... and suddenly, whhhooophhffff! The star quickly dropped from the sky as the light dissipated with barely any sound. It gently positioned itself about 200 feet away and at about eye-level with me. It was a spherical shaped craft the size of a compact car, just hovering there. It seemed like it was a very light metal and like it was such advanced technology that movies couldn't even replicate it. Or have they? I honestly wouldn't know what type of technology this was. I've never seen anything like this, nor will I probably ever again in my life! "Is this really happening to me, this is real, right?!?" Pinching myself and checking my senses and tapping my face, I thought, "Yep, I'm definitely awake." I just drove 45 minutes and this sure isn't a hallucination. This super advanced object now started to slowly rotate clockwise around to face towards me. An oval-shaped window panel with an aqua blue light emanating from within appeared, vaguely revealing one being that was flying this craft. What seemed to be two arms, a slim body and slightly larger head - maybe it was a helmet - appeared. We stared at each other for a second. I thought, "What do I do now?" A thousand more questions flashed through my mind trying to figure out the best approach to this situation. Can you imagine? You're just standing out on the street at midnight next to your car, and this thing starts to hover in front of you like a deer in headlights. You can't yell for anyone to call the authorities because this thing will be gone before people can even hear you. This was my chance to attempt to communicate with this incredibly supreme being. My mind cycled through the best options I had; "Do I bow down? No, I don't want to surrender myself or make myself seem insignificant to this being. Do I put my hands up? Run away from it? Do I run towards it?" Finally, I decided I had to do something and I pointed directly at it with my left index finger, saying, "I see you... I see you... I see you..." Once I said that, it slowly lifted up and to the left, went over and then behind a house, carefully navigated through backyards and made its way up the street past me. I was able to get one last good side view of this object and the being flying it. It had slightly longer arms than a human and was definitely thin-bodied. The craft itself had a metallic-like seamless structure and simple design. I wanted to follow it further up the street but decided I didn't want to force any further interaction, possibly causing trouble; I definitely wasn't ready to be taken away. So I got back into my car only to drive several houses down the road. They were going to know exactly where I lived if I drove straight home! As advanced as this being was, it probably already knew a lot more than that about me. At the time, I was still living with my mother and grandmother. Finally parking in the driveway I noticed my mom's light was still on so I was able to tell her what had just happened. I told her I saw a UFO drop out of the sky in front of me. She quickly hopped out of her bed, where she had been reading, and asked me, "Where?!" as if I brought a new puppy into the house. Sarcastically, I said, "Oh, they're waiting for you outside, I told them you'll be right out". "No, they're gone but it was right in front of me on the street and we were looking at each other, it was right there!" Of course, she was aware of all the other sightings I experienced and the only thing she could tell me was that there are some types of angels trying to send me a message. What could this message be? I did ask for a sign when I first saw the drifting stars, and I specifically asked for help, but could this be it? Maybe it was a way of telling me that they are coming closer and will gradually introduce themselves. I didn't know what to expect from here or what could happen next. It's a strange feeling when you think you've become a target. Why was I going through this?

My mind was so distracted and shocked from the crazy things I'd started seeing. I feared the idea that they may now be following me and could possibly abduct me. I tried to remain healthy and focused at my full-time job as a graphic designer in Norwalk. I felt it was so serious that I had to inform my manager, Shay, about what happened just in case I

didn't show up one day. Abduction would be the only reason, so there would be no surprises. She asked me if I needed to take some time off to gather myself but I told her I had to stay occupied on something else and going to work was one thing where I was able to redirect my thoughts.

During a night with the co-workers in South Norwalk Connecticut, we went to a paint and sip social bar. The theme and instruction for everyone to follow was a mermaid perched on a rock surrounded by calm water under the moonlight. Rather than a mermaid, I painted a dragon, all black holding a crystal ball, and a crisp moon behind it. On the top right corner was where I put my interpretation of that spherical phenomenal flying UFO as a little message to all and reminder of exactly what I saw.

There would be days when some would suggest that I shouldn't tell many people about these experiences because then people would think I am crazy. I understand it is something hard to believe, but the truth can be shocking. These unidentified objects were always on my mind and what could it all mean? These strange sightings happened multiple times in such a short amount of time, and they'd been slowly coming closer and closer to my line of sight. How close would they possibly get next time? Will there be a next time? I'd need to resort to better equipment to take better photos and video. For a while, I was refusing to become dependent on a smartphone device since I noticed the way it had started to distract people, especially people my age. If something happened to me, at least people knew me as the guy who was seeing aliens. Though, I would like to give these strange beings more respect and credit by referring to them as Advanced Beings. Who knows, they could very well even be the Angels or Saints we always hear about and see throughout historical art and records. Whoever they were, they seemed peaceful by their careful approach and highly intelligent based on their mind-blowing advanced technology. Who wouldn't want to fly one of those things?! I could not help myself constantly talking about something completely out of this world, that which actually existed around us. The people must know, and I am a witness.

A few weeks passed and I had come across an article about a couple from Northwest America that described these stars in exactly the same way that I had been seeing. They noticed one as they were driving home and thought it was odd the way it was moving in their direction. It seemed like it was following them home but then they forgot about it when they arrived. Later that evening, the husband heard a knock at the door and, as he went to open it, there were these slender creatures that telepathically told him not to be afraid and that he had to go with them. He responded that he wanted to let his wife know before he went with them but, as he turned around to tell her, they touched his hand and he blanked out. The next moment, he was waking up in the morning next to his wife as if it were all a dream. The wife wasn't aware that the man was even gone the rest of that night. What I gathered from this article was that they were followed home and these beings took the man, or maybe even both of them.

I seemed to be the only one noticing these stars, could they be following me personally? I became more paranoid and frightened of being taken. Since I didn't know where to find more information regarding these stars and flying objects, I thought I would prepare myself somehow in the event that I do get to meet these beings. I had this feeling, like a tingling in my head, every time they passed by in the sky, forcing me to always be attentive when they were around. Like a tap on the shoulder saying, "hey look". I thought more about how I could communicate with them. I began taking an interest in the Sumerian alphabet and learned a few basic words. Since it is one of the oldest languages we have on record, maybe they'll be using something like that to communicate. For example, in Sumerian, Earth is Kia, like the car.

One crisp and clear night, about five weeks after my last major sighting, sometime around the end of April 2011, I was walking to my car in the parking lot of a bowling alley in Fairfield, CT (I belonged to a league that played Tuesday nights) and looked up to see one of those stars and its bright light steadily drifting from the East! This star was a tad brighter than the other stars in the sky, which means it could be much closer to Earth. After what I witnessed a few weeks ago, I now knew what this thing was. Casually communicating with this flying star, I mentally projected to it, fully knowing that I couldn't actually speak to this object, but it was worth a shot: "Hey, I see you. Where are you going?" No answer. There was nothing that I could do about it, so I hopped into my car and made my way back home to Danbury. Only ten minutes into the ride, I was on the Merritt Parkway and I looked out of the driver's side window and saw what looked like that same star again! It was the same light, same speed, same direction. Slowing down the car a bit, I wanted to get a better look to be sure and indeed it was that same drifting star. The first thing I thought was, "My God, is it following me home now?!" I decided to get off the parkway an exit early to make sure it wasn't, and maybe fool it into thinking I was going somewhere else. After getting lost and using the GPS to get me back on track, I finally arrived home an hour later. I parked in the driveway, got out of the car, and as I walked to the garage I felt I had to look up and I saw that same star from earlier! It was the same brightness, altitude and speed, going right over my head. Immediately, tears came to my eyes because in my heart I felt and almost knew for certain that this was coming, this could be the sign I'd been asking for. I recalled the story of the couple who was followed home and the man who was taken and returned to his bed in the morning. I thought this was it. They now followed me home. But I wasn't ready to be taken. I was terrified of the thought that this could actually be it. These consecutive sightings of the same object in the same night don't ordinarily happen.

I went into the house and informed my mother and grandmother that I had just been followed home by one of those stars that I'd been telling them about these past few months. I told them, "If something or someone knocks on the door, let me answer it or, if I am not here tomorrow, then you know why." My mother thought she was funny by saying, "As long as they leave five-million dollars, then it's a fair trade." I wasn't happy about her response and told her so. She corrected herself: "Fine, five-hundred million." And laughed. "This is very serious!" I shouted. "And so am I" she said. My mother and I share an odd type of humor but, in this particular case, she wasn't helping my situation so I started to ask my 84-year-old grandmother about our Hungarian background. Could she know of any reason why these things are following me? I asked her, "Why is all this happening to me? I'm not ready to leave." In her very strong Hungarian accent, she told me, "You don't have to leave. Just tell them you want to stay and they will go away." She was right, what other choice did I have? I was nervous and paranoid.

My cat, Angel, was also there to comfort me and make sure I was okay. He had been my best friend for about fourteen years at that point. He was a brown, short-haired domestic tabby with black stripes. He was my buddy. He followed me everywhere and slept in my room almost every night. Back when I used to come home from school off the bus, Angel would be waiting at the rock at the corner of the street everyday. We had been through so much together and he was one of the coolest cats. I'm sure he was sensing the worry and fear I was going through. Nothing happened that night and for the rest of the week I had trouble sleeping so I kept a light on and would wake up throughout each night, checking if I was all there and if everything was the same. I felt a strong sense of being watched. It was absolute paranoia. My instincts were telling me something was coming.

The Friday of that same week, after work at 5:30pm, I had a softball practice at the beach in Norwalk. Some of my teammates knew about the stars I'd been seeing and that I was always looking out for anything unusual. I was just happy to be out there with a great group of friends to take my mind away from it all. I remember it being a very windy yet beautiful day with the sun setting behind the boat yard, bringing me a sense of peace and comfort as I stood out in the field waiting to catch a fly ball. I still had that sense of something watching me, but I was determined to remain focused and balanced, to play my best and hardest. I wanted to impress these advanced beings, if they were in fact watching me and planning to do anything to me. I was expressing my value and worth as a physical being.

I arrived back home later that evening, my body was beat up after a strenuous game and I was exhausted. I wasn't just physically exhausted, but I was mentally drained considering all of the experiences that had happened within the past few months. All the fear and paranoia had put a toll on my well-being and I just wanted to go back to living a normal life. I thought maybe a good hot shower would help and then I could try and get some sleep. Deep within my heart and soul, I attempted to send a message out to these beings that I was tired of being afraid. "Whatever it is you need to do, please let it be over with already. Please do not hurt me or take me from this Earth and my family as I still have many things to accomplish. I am not ready to go. Thank you for your mercy." From here I was basically giving permission for these beings to do what they came to do, because I have had enough. I was to let go of any tension from worrying so much. I went to bed and, this time, turned the light off and acted as if nothing would ever happen. I just wanted some sleep!

I had no recollection of having any dreams that night. Not once did I remember tossing or turning at any point. I'm typically a side sleeper but, when I awoke early that following Saturday morning, I was laying perfectly flat on my back with my arms tucked by my sides. My eyes slowly opened and I stared blankly at the ceiling. Quickly regaining consciousness, I began to sit upright in my bed with my body feeling completely numb, as if I was coming out of a deep sedation. I started moving around my arms and turning my torso to stretch a little while thinking, "Wow, I haven't slept this good in years!" I continued stretching but, as I turned to adjust my pillow, I noticed a fresh blood stain on the pillow where my head was. It was as if the blood squirted out in one thin, straight line about two inches in length. I gently touched my face, ears, and nose to see if there was any other blood but nothing was there. "I wonder where that could've come from" I pondered in a dazed state of mind. "Where's my cat?" Angel is usually the one waking me up in the morning or just always there. I didn't feel surprised to see the blood. I didn't have any strong reaction to any of these abnormalities. Completely phased-out, and I sat there just thinking, "I should get up out of bed." I swung my legs over the right side of the bed and walked to the bathroom with a smooth and gentle stroll like I was walking on air. As I began to wash my hands, I looked into the mirror and saw these razor-thin vertical lines mirroring the left and right side of my face. There were two on the forehead, one on each cheek, and two on my chin. What the heck are these!? Touching them and taking a closer look with more light, they weren't sleep lines or cat scratches. They were just beneath the top layer of my skin. They appeared as if they were a type of microscopic tube, much thinner than a needle, that left very tiny waves in the lines. They were perfectly symmetrical on my face. "Let's see what my mother says." I quickly thought to show her. Finding her in the kitchen, I asked her to tell me if she saw these markings. "Yes, I see what you're talking about, kind of. They look faded, you should go take a photo of them if you can." I'm glad she was thinking quickly because I was still amazed by what I saw. I quickly located my small digital camera, "Oh! The battery is dead, of course!" I found the backup battery, popped it into the camera and took two photos close up of my face to see if they were there, but I couldn't get it. They were already gone. My guess is that my blood started flowing through my body and cleared them away. There was my proof, but I wasn't fast enough to capture it.

Now that I was all perked up and awake, I tried putting together all that I noticed from the moment I woke up less than ten minutes ago. I thought, "Hold on, did something actually happen last night?" I had to take a moment and think about everything. I sat down on the edge of my bed and replayed each step: "The unusually fantastic sleep, the blood on my pillow and the fact that my cat wasn't there. And those strange lines that were all over my face. Something must've happened." Right at that moment, my cat Angel walked into the bedroom with worried eyes. He came right up to me, putting one paw on my knee and the other paw reaching up to my face, gracefully tapping my chin repeatedly while making an unusual meowing sound: MaaoOwWww! In all the years I've known him, I have never seen this behavior. If I had to guess, he was trying to see if I was okay and check out what happened. But this was undoubtedly unusual behavior coming from him. He must have seen something strange, which means these beings came into my room while he was there watching. If only he could talk! I picked him up and cradled him as I said, "Angel, I'm André, I'm alive and I'm okay. Thank you for telling me." Then I placed him back down and noticed his concerned and curious expression as he carefully studied me with much intent.

I had no recollection of anything that happened through the night while I slept. Maybe they worked on me at my bed, which would explain the totally random blood stain on the pillow. Quietly, I thought to myself, "Well, now what?" I wasn't flipping out or going crazy over what happened, but I acknowledged it and was grateful to be alive and breathing. I had no idea what exactly they did to me but I was very grateful to these higher forces for not doing any harm and keeping me safe. After a few minutes, I noticed there was a change in the way I felt; I had no more fear or paranoia. I felt there was a greater purpose for me and that I had to strengthen my brain and body in order to get there. Whatever it is that I had to do in order to reach greatness, I would do. There should be nothing to interfere with my purpose and growth; fear, doubt nor depression will get in the way. I felt that an upgrade and transformation must take place and that the means were out there for anyone willing to go after it. Now was the time to make those difficult and bold steps in order to gain the strength and willpower to find the path I was meant to follow. To prepare me for the coming years ahead.

A couple of days passed and I discovered some new markings on my body. While washing my face, I felt on both temples of my head what seemed like pimples or boils but they weren't quite exactly those. I couldn't pop them, as they were under a deeper layer of skin and something was telling me to leave them alone. Then, in the shower, I found a third bump, just like the other two, in the center of my pubic area. The locations of these anomalies were right over significant areas on my body. My only assumption was that whatever has been injected or inserted into me from the other night must be healing or coming out. These markings can be seen in the video I recorded about three

April 2011 - Just before the visit

December 2011

months after the visit, which remained as scars for at least another year.

From that night on, aside from work, all of my daily thoughts, goals, and duties transitioned from "what's out there and who is following me?" into, "why do I walk and talk the way that I do?" I observed how others moved about this earth in their own unique and creative ways. It is all by design, and every person is their own creator of the way they have designed themselves. What they have learned and how they've adapted to their upbringings and surroundings. As I self reflected who and where am I at this point in time, I thought to myself, "How can I improve the function of this body?" By becoming more aware of my surroundings and what is possible and accessible to me, questions were flooding my mind and I was unintentionally absorbing information from all around me. Analyzing the current state of human civilization, like infrastructure, technology, ecology, agriculture, medicine, sociology, etc... It was as if I was collecting data for these advanced beings, and using my eyes and body to observe the world. Most of these concepts were already familiar to me so why would I need to question them now? What I felt was happening was that these beings had connected to my mind and body, assisting to improve the way I functioned and showing me what I am truly capable of as a human. This was all at the cost of using my body as a satellite or guide for them to transmit information about our world and the interactions I experienced through my eyes. I was beginning to get the sense we were all being watched, as if we are one big scientific experiment. They were observing how far we had evolved and where we have consciously ascended to.

I was consistently waking up at 7 O'clock in the morning and going back to sleep at 10 O'clock in the evening. I didn't need an alarm clock to wake and I always had everything prepared for the following day. Once my eyes opened, I was attentive and ready to go like it was a holiday - everyday. There was no hesitation or procrastination, like we sometimes feel in the mornings. I felt there was a connection as if I was being pushed and motivated like a machine or avatar, which continued this way for the next three years. I never had any headaches or got sick, I had very few to no allergies and my asthma was doing extremely well.

My body was going with the flow and I did not resist, nor was I able to. It was like I was a plane on autopilot and I could only watch from the cockpit. I was in the vessel but there was a higher force in command. Since all my fear was completely gone, I trusted they were taking my body through a process and journey leading me to a higher purpose that I was meant for, but I had no clue where or what that was going to be just yet. I was fully focused on my immediate priorities. My daily routines and chores became more efficient, including health and hygiene practices as well as meditation.

The change in facial expression when renewing the drivers license. Stunned.

No time was wasted when showering. Rinsing from head to toe and wiping away the excess water from the arms, torso, back and legs when finished before drying the rest with a towel. There is something about moving and charging the energy in the body by doing this, and I did this the same exact way every time. I made sure to floss every single night, rather than the few times a week like I used to. Then watching myself in the mirror, looking deeply into the pupils of my eyes with the sense of something staring right back through my own eyes. "I see you." I would always say. I felt that these beings were absorbing every moment, meeting, interaction or scenario, every bit of information and experience, through my body and senses.

I wanted to learn so much more. From a new found perspective, I realized that I was far behind and had to catch up to where this mind and body should be: a much more capable body than I, André, had been utilizing. I needed to improve all the areas of my life that make me the best of who I am and why I am here now. Why are we here now? After all, this was my original request to these "star beings".

I remember running into some old friends from high school who had invited me to a house party in the Cove area of Stamford. While everyone else was talking and having a good time, I couldn't really participate much except by having a beer and listening. Wandering slowly into the backyard, standing by the tiki torches, I had the urge to look up into the night sky and ask, "Why am I supposed to be here?" This was the least likely place I would've chosen to be. Directing these thoughts somewhere above, one of those stars suddenly appeared! One friend saw that I was staring into space and shouted, "Yo, Dre, what're you lookin' at over there?" Then, he slowly walked over to where I was standing and looked up. I said to him, "There's one of those stars, oh wait, now there's actually two! Do you see them?" As he adjusted his glasses and started to really focus, he saw one: "There goes one, oh I see the other. Whoa, there's three of them! What is that?" I said, "That is them, the stars that have been following me. I'm glad you can see them." I replied. Then a few others from the party were curious and came over to look. One would say "That's a satellite" or call it a shooting star. They all had gone through the list of possibilities just as I had when I first started seeing them. There were a total of five of them swarming above us like bees. Everyone was puzzled or startled and mumbled things like: "What the #@!% is that?!" Then they began playing with the fire from the tiki torches and dancing around the backyard. It was then that I recognized that these beings just wanted to observe my friends, as with every other situation they had guided me into. It was all part of their data collection and analysis of the way humans of my age group interact with one another; what humans have become, evolved into, and where they might be going from here.

Many occasions I would spend time with a group of guys in a part of Stamford we refer to as Jahmrock. These were very down to earth, creative and spiritual people. It was one place I did not feel so judged or made fun of. We recognized and accepted that everyone had different backgrounds and upbringings and didn't make judgments based on color, but on the "vibe" and soul we each carry. It was really our love for music that brought everyone together at Jahmrock, particularly on Sundays when we shared traditional Jamaican BBQ. There, I met two brothers, Zhibez and Ralo, who had built their own music stations and studio because that was their outlet for expressing their experiences on this earth in hopes that it can, in some way, make a positive impact on the world. As I was connected to these higher-beings, they were impressed with this group of guys, as they should be. They always kept it real and straight forward, while honoring their connections to the most high.

On some nights when hanging out in Jahmrock, I would have the chance to point out a couple of these stars to the brothers who, sure enough, saw them as well. We would try to imagine the possibilities of what they could be. Since I felt some sort of connection with these beings above, specifically this one star I was pointing out, I told them, "Watch the light from that star go out" and right then, the light actually went completely out! "Woah!" They were all shocked to see that, but then I said, "Wait.. Wait... It will come back", and the light reappeared and flew past us heading southeast. This sort of thing doesn't just happen to anyone, and what are the chances that I could predict something like this? I believe it was the phenomenal connection I had with them that allowed me to predict the light pattern, and I was able to show these humble gentlemen that I wasn't making any of this up.

Zhibez' cousin Groove was always skeptical of this experience I've had, and he walks over to me and says in his debonair voice, "Dre, you're not the chosen one, come on now." I never claimed to be the chosen one, and I'm pretty sure this has happened to hundreds of other people who may not have realized it, but I was one who did ask for help and guidance from some higher forces. These gentlemen have been friends or acquaintances of mine since before this whole visitation happened to me and they can also say that I wasn't exactly my usual self. Zhibez would recall that I often seemed "spaced out" during these few years. In other words, rather than the average small-talk, I was more concerned with the stars and higher powers existing around us. Whenever I was outside, I was constantly drawn to the stars.

Five months go by with each day pushing myself to learn and accomplish as much as I can. I had reached a point where it was time for the next phase but I didn't know what to expect. My father had been battling cancer for a few years and we weren't sure how much time he had left. It was crucial that I spend as much time as I could with him. It was hard for my dad to process the experiences I went through, but knew I wouldn't just be making something like this up. He opened himself to the idea that these beings were possibly angels to watch after me because he was being taken away from me way too soon. He requested that I ask if they could help him in some way. That night I pleaded for mercy and to allow him the chance for more time but the following night he passed on and his soul had left his body.

My father, Claude, was my mentor and my role model. I looked up to him in a way that I was looking at myself in the future because I am his first and only blood son. According to him, our family history is very important and significant, and I believe him about it to this day. We are descended from a family line, passing down the highest code of morals and traditions with integrity, equality, loyalty and honor. At least that's what he believed and I do hold myself to such standards. Our ancestors consisted of Knights and protectors of King Louis XIV, Ambassadors to Napoleon Bonaparte, military leaders, as well as wine and brandy farmers and workers in the French steel industry. Claude was a very impressive and intimidating French businessman who always had high expectations of me. Even though I did not go into business, medicine or law like he had wanted, I would try to achieve great things through art and innovation (which he felt was a "dead end" career). There comes a time in everyone's life where we must face the reality of losing someone we love dearly, and though my father was a tough and strict man, I always loved and looked up to him as difficult as it was to impress him. Life is precious, and should not be taken for granted. How tough it was to lose a man that meant so much in my life. He was the compass to the entire family! Deep inside, I felt as if these beings were there to help me through the loss of my father. With all the recent adjustments happening to me, from the shock and paranoia of experiencing these advanced beings, I naturally became numb.

The holiday season was even more difficult to feel any happiness or sadness. As my sister Claudia and I would go through our father's paperwork and photographs, my emotions were muted on the outside. Internally, I placed most of the pain and sadness into a compartment of my brain and soul that allowed me to move forward, but I held onto all the great moments and lessons. Learning from this pain and the great lessons he taught me, I would slowly, day by day, see the man I am soon to become.

Shortly after he passed, my mother called my cell phone and said that while she was on her computer at her home in

Danbury, there was a deck of cards with a rubber band around them which had snapped, knocking all the photographs off the shelf except for the photo of him and I when I was just a young boy. The cards had fallen on the floor with only one card facing up. The ace of hearts. In regards to the tarot, Ace of hearts represents an endless stream of healing through love. I had no idea what that could've meant at the time but I took it as a positive sign that my father was still around watching over us.

Dealing with his loss wasn't easy for sure, after everything I have endured in the past year. Why did this all have to happen now? At the beginning of the year, I had begun to take action in my life and then everything came crashing down. The only thing that remained was a hollowed out core of my body. The distress I felt in me would naturally make me want to curl up and weep, but with this recent connection/ visitation I was able to keep my head up and make it through these tough times. I was also fortunate enough to have a good support group around me like the Pastor and his family, my coworkers, teammates or the fellows from Jahmrock. They had helped to keep me emotionally strong. Also having a job doing what I love, creating graphics and imagery for a large tradeshow company, helped; it was the one place I didn't have to think about my own problems all day. My skills were rapidly developing pertaining to my craft of design. It was very exciting to have had this opportunity. To see my hard work displayed in two of the largest convention centers in the nation, I am humbly grateful for all of the blessings around me in life. By recognizing my strengths, it allowed me to have the focus and vision to move toward higher goals and dreams, most importantly: purpose!

I had become restless, constantly in deep and focused thought. Now that I established a daily work schedule, I simply mapped out on paper what I do each day and filled up any free time with the things that I felt most compelled to do. This energy and determination to propel me forward didn't feel like it was solely 100% me. There was something that was pushing me harder and faster, moving myself either by walking, running, jogging or driving with more precision, intent and caution.

Every evening after work consisted of coming home to prepare the same plain pasta or salad while watching Jeopardy and practicing Hungarian with my mom and grandma. Then, slowly working on sit-ups and pushups in the driveway while observing the occasional star beings float by, honoring and thanking them for this guided and focused drive that I did not quite have before. With this laser focus and work ethic, it was time to put it to good use and apply it in the areas of my life that must be improved.

I have always been athletic but I was nowhere near what I could be and the gym wasn't giving me the challenge that I needed. Coincidentally, through social media, I came across someone I recognized from my old high school but never formally met before. He goes by many names but I'll refer to him as Three-sixty. He was just starting out his own training business and I immediately messaged him about his training services. This couldn't have been more perfect timing. He saw the willingness and motivation I had, and it wasn't long before he saw the effort I made to push myself further and further. I didn't have any other choice. This is what I was guided to do. It almost felt like these beings

led me to this vibrant, young, tall, athletic man. Three-sixty and I started by meeting a few days a week at 6 o'clock in the morning. This gave me time to work out and have a shower before going to work by 9 AM. The determination to wake up early fueled me to accomplish even more through the day, both during work and after! When it was time to work out, Three-Sixty pushed me to levels I never could have imagined reaching. He was like the commander and I was a soldier in training. We shared many discussions about how life and humans could be. Not only did this man transform me physically, but mentally passed along knowledge, wisdom and understanding.

"There are three's in many things and it all circles back around." This was his thing, this is why we call him Three-Sixty. I would like to share with you something that he has taught me: we must learn to accept, rather than expect. It's easier to love someone when you accept him or her as they are. This piece of advice has opened my eyes to treat many interactions with better intentions. Humans are known to make mistakes, and at times their behaviors don't align with what we would expect them to be. By accepting the way someone speaks, acts or behaves just as who they choose to be, we may find a better understanding and connection to where they are coming from or going. This doesn't mean that we should tolerate someone who is immoral or cruel, but by accepting the way someone decides to live their life is their choice and not ours. This type of energy does not do anyone any good, like judgement.

Now that I had adjusted to this new schedule, it was time to add something to my weekends. I thought back to when I first moved to Danbury, not too long before this. I had moved right near the local airport and I've always wanted to learn how to fly.

Just landed after the introductory flight at KDXR
Danbury Airport with Arrow Aviation

One of my French ancestors, Frederick d'Esterno had written a pamphlet on the flight of birds which had a major influence on the future of aviation. So I felt that it was in my blood to give it a chance and, as all of my fear had been removed after my last visitation, I did not hesitate to begin taking pilot lessons. Again, this was with the guidance and push from these higher-beings connected to me, as I would have never just decided on a whim to do something this bold. Prior to all these changes happening in my life, I never would have imagined deciding to learn how to fly, but there was an intense desire in obtaining this most vital information to benefit my growth as a human. Was this a test to find what I'm capable of? One hour of flying every other weekend, and completing ground aviation classes, I was already getting comfortable taking in all this new information. Cliff, the flight instructor told me what normally takes 30 hours to learn, I achieved in only 10 at the time. I'll never forget the look on his face when I landed the plane perfectly smooth for the first time. He was genuinely stunned and so was I! Only, it felt like I had done this before.

It's been several months since the last visitation, and I have become more efficient in my daily priorities and chores, for now. As I've been so diligent and productive each day, I was sharp and squared away with everything that needed to be taken care of. One night, once again finding myself laying flat in my bed, still at my moms' house, in the same bed where I was visited previously, my eyes opened up to my dark bedroom where I can see my wall closet beyond my feet and the box television to the left next to the window. But instead of darkness, I saw a light-blue aqueous glowing bubble the size of a soccer ball. It was some organic, liquidy ball similar to when you blow a bubble and there were other little bubbles morphing around it. It was just floating over the left side of my hip. I couldn't tell what it was. I tried to reach for it but I couldn't move my arms. My body seemed to be in a state of paralysis. My eyes began adjusting to the rest of the room and I noticed a silhouette of a figure standing to the right of me, beside the bed, making swift hand motions over my abdomen and hips. "Who is this? Who are you!" I began questioning, but the words weren't coming out of my mouth because I was in a deep paralysis! "Do I trust this? What is happening?" This wasn't a dream and I have to get out of this. I started grunting inwardly, making it louder in my mind, getting louder and louder and the next thing I knew the blue bubble popped and faded away. Then, the figure stood up into a standing position and teleported up and away like a digital glitch, in a vertically upward motion. Sswwffff! The moment this figure disappeared, I quickly sat up in my bed and looked around the room saying, "Wow that just really happened, that wasn't a dream! It couldn't have been!" Was I not supposed to interrupt their interaction with me? Whoops. I just wanted to communicate and maybe understand why this is all happening to me. After all, this is my body and I ought to know what's happening to it.

My own doctor wouldn't believe the series of events that I had gone through earlier in the year. He asked if I had any proof. Dentists have said that my teeth are grinded down so much, they resemble those of an 80-year-old man, but my doctor couldn't confirm that the state of my teeth was connected in any way. Other than the video clip I took explaining what I was going through, there was no easy way for me to predict when something was to happen to me where I would be waiting with a camera in hand. I ought to be recognized as an honest, equal human being who's giving a truthful statement in order to find an answer to why this is happening and what changed in my body, or if any foreign objects were implanted, or if anything else had physically changed. There's only a strange, questionable, small nub from a scar by my bum hole that is fairly sensitive - like a paper cut.

The doctor referred me to a psychiatrist and, after 45 minutes of explaining my entire story, all they could offer was to zap my brain and put me on medications, which I obviously refused. Furthermore, I then had to pay them $125 for that session out of my pocket because insurance wouldn't cover the co-pay. It was difficult finding professional help as this is all a very strange and specific experience and I know I'm not crazy. Actually, I am quite sharp, healthy, honest, respectful and thoughtful as I've always been. When it comes down to something like this, there is no honest hope for me to find professional help. The only thing to do is to learn as much as possible and improve my life in every way.

Who I am now isn't exactly the same man as from a year ago and it is part of me to always try to share my story with everyone. We can only imagine or assume there would be other life out there. Are they looking after us? No one really knows. Having this discussion of advanced beings or aliens is the quickest way to make people think you are crazy. If only I was able to record that star dissent from the sky which hovered two hundred feet away from me, or videotape myself while sleeping, that would've been all the proof I needed. Unfortunately, I wasn't prepared for something that intense and all I have is my honest, truthful word from my sane mind and sharp eyes. My cat, Angel, was a witness but what could he say: "Meow!"?

As I maintained focus on my daily missions and all of my interactions with people, I felt numb to most human emotional connections. Almost robotic in a way, I was gathering some beneficial evidence about the way we are to the earth and each other. I made a habit of sharing jokes with coworkers because it seemed to boost the morale and vibration of the environment. There was a part of me that would not react so easily to life's disturbances or aggravations, like traffic. By leaving early and factoring in a little extra time, I follow a safe pace on the road and not worry about what other drivers are doing, and I manage to get to my destination safely and stress-free. But there are a lot of crazy drivers out there. I've made the observation that the way a person drives can determine a lot about their behavior or character. With just a little patience and courtesy towards each other, the roads would be a lot safer and kinder! That goes for watching out for animals crossing the road! Many of us are caught up in distractions of our minds and tend to neglect consideration for others around us. We are habit-forming creatures, and through the power of the mind we can break bad habits. Is the digital and technological world separating us from our natural states of mind and existence?

After these visitations, I never got sick nor did I have any headaches. My yearly physicals were exactly the same results for three years, which shows how consistent fitness and proper nutrition have helped me dramatically. I have managed to accomplish so much more in these last few years than in the previous ten years. But then, one day, I woke up feeling different. Normally, I would jump out of bed and get going on the first chore or mission but I didn't have that same energetic force, urgency or drive that was pushing and guiding me for the last three years. I thought, "Wow, so I have to do all that on my own now?!" I was back in the pilot seat of my own ship, except now there were a few more gauges and buttons. I was still working out on the beach with Three-sixty in the early mornings before work, taking some salsa classes, studying German and Spanish, checking in on friends and their pets, and establishing myself as a graphic artist and photographer. All I could think was that these beings brought me to the path I had prayed for, and now it was time for me to prove that I can command and control myself, but without that external push. There is a way to structure every hour of the day so that each minute moves towards something we would like to achieve; by dedicating our focus to that one specific thing without distractions. By applying yourself little by little each day, you eventually will gain or achieve so much in only one month! It's incredible how we can make so many things happen to improve our world and lives and yet we are so distracted and consumed by other things in our world today.

For the three years after the visitations began, I did my best to keep up with what many would consider a busy schedule. I had moved out of my mom's house and found a great second floor rental apartment in Norwalk Connecticut. The deck from the bedroom overlooked the boating docks out towards Long Island Sound and Veterans Park, where the famous Oyster Festival is held every year. I had to sacrifice pilot lessons to afford to live there. It was only a four-minute bike ride to work, right by the water, and it couldn't have been a more perfect set up for me. I now had the space to find myself again, regaining the senses of the human experience and the control of my own thoughts and emotions, not being led by some outside force that guided every movement. I began to feel my humanness and my emotions naturally occur more frequently again, like laughing or crying.

I hadn't directly experienced anything more from the higher-beings, but those drifting stars were there from time to time, and they still are. I now called them either the Annunaki or Pleiadians. They seemed to be the closest descriptions to what I had been seeing. I'd also heard the word Anu from Hungarian, which means Grandmother.

During my time in the Norwalk apartment, I recalled some strange things in the environment. One day I had come home from work and took a moment to stand out on the deck looking over the docks. It was a beautiful clear day. I looked closer and there were straight horizontal lines of transparent vapor, creating a grid going over me and out

1. Vapor lines present   2. Moon Appears and vapor lines gone

towards Long Island. I tried taking the best photo I could to capture these vapor lines but was very tricky as they were almost transparent.

Then, about 45 minutes later, I went to look outside from the deck once again and the lines of vapor had disappeared and the moon was suddenly there. It wasn't there a few minutes before, it had suddenly come out of nowhere! You'll notice in the second photo a clear blue sky with the moon now up high above the tree.

That same night, I was on the phone with a friend while sitting out on the deck explaining everything that I saw earlier that day. As my friend spoke, I looked up to the sky and saw that there were these ripples or waves flowing through the air from a satellite into my phone - what could've been data or digital sound bytes of some sort. They were synchronized with everything my friend said over the phone. How could I be seeing this? Was the humidity or density of the atmosphere creating a perfect condition to physically see sound/data waves ripple through the air? Did it relate to the vapor grid that I saw earlier? I really wasn't seeing the world the same way as I used to before the visitation. There's apparently much more to this life than what we know from history books and what we've been taught or shown. Even our eyes are unable to see every sound, x-ray or frequency. We must be more open-minded to the fathomless technologies that may actually exist.

After a while, I began actively looking to invest in a home instead of renting a place. It was impossible to compete with other builders and investors, as I was just a first time home buyer, but after two years of searching along the coast from Stamford to Fairfield my longtime awesome neighbors next-door to my mother's house in Danbury were looking to sell their home and were willing to do a private sale with me. They knew it was a perfect situation to be next-door to my mother and grandmother, to be able to help them out while keeping the house in great shape like Mammy and their Poppa Tom had done for years. It was a true test to keep up with all the responsibilities of maintaining a house, being my own physical trainer and commuting to and from work. Buying their house seemed almost like something out of a fairytale-come-true. If only it were as easy as it sounded. Though, what other choice did I have? It was the hand I was dealt, and this must have been what the Ace of Hearts was all about.

After a year of living in my first home, it was now 2016 and I was in the same neighborhood I spent half of my life. I had to make another step further and at this point, any idea seemed to be too risky. Something about the world and economy seemed to be changing and I had to make a move. I decided to leave my full-time job and work as a freelance graphic designer and photographer. At the same time, I wanted to write out this memoir to let as many readers and listeners around the world know what I have experienced, and to know that everything is true. If you have had similar or identical experiences just know that you are not alone. There are countless others who probably hadn't noticed any visitations but perhaps had a sudden change in routine, habit, focus and intentions, maybe even developed a heightened sense of awareness.

I was getting so caught up in the distractions of life; the times clients have you do months of work and decide not to pay, the times police enforcers wrongfully accuse you of things you haven't committed. It was hard to find focus, as my thoughts were scattered everywhere. If only these higher-beings could come back to help me.

I wondered: "What's missing from this equation of all that I'm doing, or the lack of doing anything?" Strolling around my street at night, I looked to the stars for guidance as I always have. They were all shining above me silently listening to everything I was prepared to say from my heart. There was space for a dog in my life, and now would be a good time to train them and raise them to be a great little helper for the future. Dogs also need structure and consistency, which would be just the thing to help me incorporate more structure and consistency back into my own daily habits and rhythm. What would be the best breed for me? I prefer larger dogs and they would have to be hypoallergenic. They'd also have to be active and sturdy. I listed to myself all of my favorite breeds: Pitbulls, Shepherds, Rottweilers and Boxers. The desire for a dog became a prayer: "Please, Lords of the heavens may you find me a dog that will be a good earth guide and protector."

Not even two weeks later, sometime around midnight, I noticed an ad from a friend on social media to adopt a Boxer-

Bull puppy. Thank you Shara! There were nine in the litter of pups. Without hesitation, I applied and I was lucky and blessed enough to have the honor of adopting one of those nine puppies! She was one of the three I had my eyes set on.

I named her Bella-Rose because she was not only beautiful but she loved playing with bells as a puppy. Bella-Rose the Boxer-Bull from Bridgeport. A very intuitive friend I had made from the west-coast also recommended this be her name and though its a very common one, it couldn't have been more appropriate and fitting.

Now, I had to come up with a new schedule to accommodate her needs. Food, exercise and sleep. Bringing her to the dog park in the mornings was the best thing for the both of us. I was learning a lot about the behaviors and instincts of dogs and their owners, as well. Playing with her at the park also gave me the opportunity to gather my thoughts and plan out my day. While planning and pondering, I would notice the increase in chem-trails being sprayed into the sky and I would think, "Those really don't look natural."

I would overhear people at the park discussing politics and what the news and media channels have told them as they display anger and hate over certain politicians. It felt like a dense, lower vibrational frequency was affecting the population. While almost every news channel had been making Donald J. Trump look and sound like a bad man, it made me wonder why they were so condescending toward him. There were presidents that everyone had doubts about in the past but never has a president been so ridiculed and hit so hard by the media. I was willing and open to everyone's thoughts and perspectives on this subject. A long-time friend of mine has been very insightful about current events and how things evolve and escalated to where we are today. Introducing me to the other side of news and media, I was starting to see a broader perspective which was intensely complex. If only I had learned about all this sooner, though I guess it just wasn't the right time until now. My friend, Ritz, would tell me that this could be the greatest movie playing right before our eyes. Indeed, there have been things going on right under our noses and behind the curtains. With cover-ups and corruption and the mass majority seeming blind or helpless against the tyrannical giants that are actually pulling the strings over major developments around the world. Soon, I began having heated debates with my own family. It couldn't have been a better time than to have my dog Bella with me to remind me of love and compassion.

Fast forward to January 2019, I started having super sharp pains in my chest next to my heart and my asthma began to irritate me. It felt like there was a growth on my heart, pinching a nerve. I went to the doctor and was prescribed two kinds of asthma medications, one of which was giving bad side effects. I researched it on the Internet, and the side effects were all lethal symptoms! So I discontinued taking that and chose to suffer while trying to self-heal, naturally. All I could do was take Bella to the park for her run and get back home to rest. Any movement or leaning over my laptop was painful for me. I couldn't exercise or complete any work without worrying if I was going to have a stroke or heart attack. After a few doctor visits, chest x-rays and blood tests, they had no idea what was going on. I was "perfectly healthy" they said. At this point, I was losing faith in the medical field. After all, they practice with medicines they don't always know everything about. "What's happening to me now?" All I could think was that maybe those unfamiliar advanced beings placed something into my body that would be foreign to our nature or technology when they operated on me. Before something could potentially happen to me, I couldn't waste anymore time. I had to finally write this memoir about their visitations and the motivation, guidance and dedication they have taught me. If something were to happen to me, then no one will ever know this story. I had to get away from everything to concentrate, so Bella and I took a ride to Miami for a week during the Fall of 2019. I was only able to write and rewrite the introduction several times before I had to come back to Connecticut and to all of my responsibilities. Bills were piling up, savings were running low and the stress was building up over all the drama and frustrations going on around me and the world.

After a few weeks, the pain in my chest became so unbearable that I had to go to the emergency room on October 18, 2019. The EKG was indicating a hesitation in my heartbeat but, even after two more chest x-rays and more blood tests, they had no clue what it was. "You're perfectly healthy," the doctor said. "If I was perfectly healthy, then why am I here?" I replied. "There's obviously something wrong here." Their only diagnosis was acute pericarditis, which is an inflammation around the casing of the heart. They wanted to prescribe me anti-inflammatory drugs and pain medication but, because of my lack of trust in modern medicine, I refused it. I decided to take turmeric, more vitamins, ginkgo biloba and CBD oil. A few weeks later, I realized that I had gone to the hospital on the anniversary of my father's passing. That was a surprising coincidence, could it mean something significant? Each day and night, I prayed in hopes that my angels or extraterrestrial beings were still listening. "Heavenly mothers and fathers from above, I'm calling to you for help and guidance. Please heal this pain in my chest so that I can continue my path and deliver this book and your lessons to the world. I deeply value everything you have shown me and am grateful to be where I am today." When I do these types of prayer, they come from the deepest part of my heart and soul. I knew that I had to be patient, to respect and gently care for my body. Stretching, and going on hikes around Connecticut with Bella was all I could do other than a little yard work. A couple weeks later, I was stretching in my backyard and, as I slowly reached up to the sky and opened up my torso, it felt like a needle in my heart was pulled out. "Was that a good thing?" I proceeded carefully throughout the rest of the day, moving around easier and actually feeling more comfortable working on my laptop again. It was right around this time that a former coworker had contacted me about a big project for the tradeshows I used to work with. This was a major surprise and a blessing in a disguise! It brought back my creativity that had been dormant the past year while suffering from the chest pains.

By taking Bella Rose to the park just about every day around the same time, she made two Labradoodle friends, Jake and Charlie Brown. There's something about big fluffy dogs that Bella just loves. A man named Eduardo would bring them to the park. He is a super kind, friendly and peaceful Italian American and we connected easily. Have you ever met someone who's energy was very calming and welcoming? People I've met have said that about me, but now I understand what they mean. We would be able to freely discuss things going on with ourselves and the world, without judging the other. That's simply how life should be everywhere, shouldn't it? We have become so comfortable in a system provided for us all to function together in society that we may have forgotten the true laws of mankind indicated in the many teachings and wisdom from all the great men and women throughout history, not only Jesus, Buddha, Allah, but however or whomever we are connected to (unless there are some of us who have been disconnected? Or even reprogrammed?). Not only was this book to share the things I have physically seen on this earth and experienced, which almost felt like being an avatar of the divine, but to remind ourselves of our true existence where we've seemed to have lost touch; we are burning our eyes while drowning ourselves in mobile devices and television screens, not really having to face people on a daily basis so we show everyone on social media what we want them to think we are. There are even filters now for the temporary sensation we could actually be different then who we truly are.

It is now April 2020, the nine year anniversary since my first life changing visitation. The entire world is on lock-down over a supposed virus which proves to be not much worse than the flu, and we now have to wear facemasks to "flatten the curve" as they say. Most news channels are creating panic and fear, showing false images to the masses, with data and statistics that aren't equating to a consistent truth. We are not supposed to leave our homes except for only the essentials. There are no more sports, the parks, movie theaters and salons are all closed, and no one is to have large gatherings. The world seems almost frozen, giving the earth a chance to heal itself and time for everyone to self evaluate and reflect on the true value of life and freedoms of humankind. So while we have this time, it allowed me to share my entire message with you all, with the intent and hope that everyone keeps an open mind to the very idea that such advanced technologies and humanoid-like beings can exist. I can't say they gave me supernatural powers or anything so unrealistic, but it was almost like they knew exactly how to rewire my brain to function at maximum capacity, showing me the focus, dedication, and progress we are all capable of if we really utilize our minds, bodies and time wisely. We need to nourish our vessels according to our blood type, though there is always that one thing we crave that could throw us off-balance. So we have to exercise, meditate and stretch to keep us balanced. Balance is very important if you want to have the best functioning body. All it takes is that one misleading step to throw it all off. However, anyone can move their body around; it is more about connecting your brain to the body. It requires determination, motivation and inspiration.

It all comes down to mastering the mind. But the mind is the tricky part to master. It is the computer of the whole operation. The heart is the battery, the lungs are the intake/exhaust, food is obviously the fuel. Despite all the toxins we take in from our surroundings, food, water, air, sound, radiation, and vibrations of many forms, the body can heal itself with a sharpened and connected brain. Through this journey I have found Sungazing to be a tremendous benefit and purification of the pineal gland, what is also known as

Cummings Beach Sunrise, Stamford Connecticut - 2020

the third eye, by decalcifying it with the warm light of the sun rays. The pineal gland is a vital organ that connects both hemispheres of the brain. The best and only times you should Sungaze is when the sun is at its lowest point on the horizon, either sunrise or sunset. It has many benefits like increasing your energy, decreasing your appetite, improving the sleep cycle, and for some it has helped improve their eyesight! The best part is that it's one of the most beautiful things to see wherever you are in the world.

Once the brain is fully connected, it can align with our heart energy center. This is our power source. What makes us so unique is the power of love. Emotions can get the best of us sometimes, but when we operate from a place of love, we find ourselves willing to do whatever it takes for who or what we love and cherish. The greatest obstacles are not always tasks, but our fears. Fear of rejection or failure. Fear of the unknown with time ticking away. Fear of the future. We are here now with the opportunity to make it a better tomorrow and for future generations. Love and honor ourselves while helping, respecting and appreciating others. Love and honor the lives we have been given, and be the better example to the younger generations. Continue to find purpose in all things you love to do. Not everyone will be just like you but we must respect and be considerate of every living creature.

Through patience, we give ourselves a chance to make better choices. We are in a phase of evolution and there's a very good chance these advanced beings may decide to reveal themselves soon. We all have to get along with each other first, black, white, brown, yellow — before we can be introduced to the purples, greens, blues and greys. Now is the time to take back control of your own body and mind, for we all must be well prepared for what comes ahead, whatever that may be. There are days where I still see these drifting stars, and in my heart I know something greater and more advanced is out there. I pray each night for them to come back and finally introduce themselves but until that day comes I strive to make the world better wherever and however I can. I pray that everyone is able to pursue their dreams and achieve their goals if they are with good intention. Keep a clear and open mind, a good heart, and healthy body. There are no undo or reverse buttons in life and we must face any and all consequences of our actions and choices. We mustn't feel the need to react or respond right away. All it takes is a few seconds to be conscious of our surroundings, while keeping in mind that we all may be under a microscope by some extremely higher force, and it is up to us; we can show them how great we can be.

I am for your dreams, continue to fight for your human rights and sovereignty, and continue to love and respect everything that gives and has life.

I love you all, with love & light.

TESTIMONIALS

The following are a few testimonies acknowledging that such an event took place as they noticed a change in the way about Andre's life.

"I've known Andre for over 20+ years now and also lived with him as a roommate of mine during his experiences. Over the course of his experiences from the beginning when seeing his first sky beings, things changed drastically in his behavior. From constantly looking to the sky at night and day to the change in his behavior as in seeming to be more focused on projects as well as his all around efficiency with all aspects of his life. I also personally observed the markings he received in the night without any rational reasoning. Definitely something not everyone goes through." – J.N.

"I have known Andre for many years and consider him to be a great mind and creative beyond his own knowledge. With that being said I have always considered these stories of alien abduction and testing as something made up in his mind, l hated when he told this story. Especially when he tells it to strangers. One thing I can't deny is the accuracy every time I hear it, the story always remained the same so there must be some truth behind it. All I want to know is why he would risk embarrassing himself. – O.G.

"I have been friends with Dre for nearly a decade and have always known him to be a passionate, inquisitive, creative, and honest person. I did notice a change in him during this time, especially around the time of his father's passing, but I also knew this change was a result of experiencing something not many can relate to. He asks you to ponder unanswered questions about life and this world. He's open-minded about spirituality and the possibilities of the universe. Ultimately, I think his message of love and acceptance is one worth embracing." – S.T.

"There was a transformation in Andre, on many levels. He became more health conscious, started to get up real early for boot camp, changed his diet to healthier veggie choices. Andre started to focus, it seemed as if he was empowered, emboldened, gradually transforming himself. His self confidence grew. His creative abilities as a designer blossomed. I noticed a change in his photography style as well. I see his pictures that are serene and calm but have a sense of discovery to them. His awareness was heightened, became telepathic in a sense. He would finish my sentences or would answer my question as I was just thinking about asking him. 'They are guiding me, showing me' he would sometimes say. I was glad he found a peace, a meaning and strength in who he is and was in character. I know this as only a mother's bond with her child can. They are still out there, he still sees them." – F.M.

CPSIA information can be obtained
at www.ICGtesting.com
Printed in the USA
BVHW092147220221
600778BV00008B/1005